Andrew R. Barron

Portland Cement in the Energy Industry

Portland Cement in the Energy Industry

Andrew R. Barron

MiDAS Green Innovations
2020

Cover image © 2017 by luismmolina

First Printing: 2020

ISBN 978-1-8380085-3-6

MiDAS Green Innovation, Ltd
Swansea, SA1 8RD, UK

www.midasgreeninnovation.com

Dedication

To my wife, Merrie, for all your support, friendship and love.

'It's an adventure'

Contents

Acknowledgements

I would like to thank Dr. Lewis Norman, formally of Halliburton Energy Services, for his friendship, scientific collaboration, and guidance.

Three former colleagues Dr. Max Bishop, Dr. Corina Lupu, and Dr. Chris Edwards are acknowledged for their work on cement while in my research group. Only two of them work in the oil industry now, since Max became a lawyer.

Last, but not least, I would like to thank my wife, Merrie, for putting up with me during the Editing of this book while 'social distancing' during the COVID-19 pandemic.

Preface

Portland Cement (or simply "cement") is the single most commonly used building material in the world today. Applications relating to the energy services industry are the primary focus of this work. In particular, the book is aimed at providing a primer for engineers in the oil industry.

A knowledge of cement chemistry and the setting characteristics as well as strength development and retarder kinetics is of crucial importance to a successful "cement job". This was nowhere more obvious when it was decisions over cement stability that ultimately led to the deadly explosion on the Deepwater Horizon rig on 20 April 2010, while drilling at the Macondo Prospect.

Chapter 1: An Introduction to Portland Cement

Portland Cement (or simply cement) is the single most commonly used building material in the world today. The manufacture of cement in the United States was in 2018 an 87 million metric ton, $9.8 billion industry (Figure 1.1). Worldwide production is at present in excess of 3 billion metric tons per year, with China accounting for 2.2 billion tons.

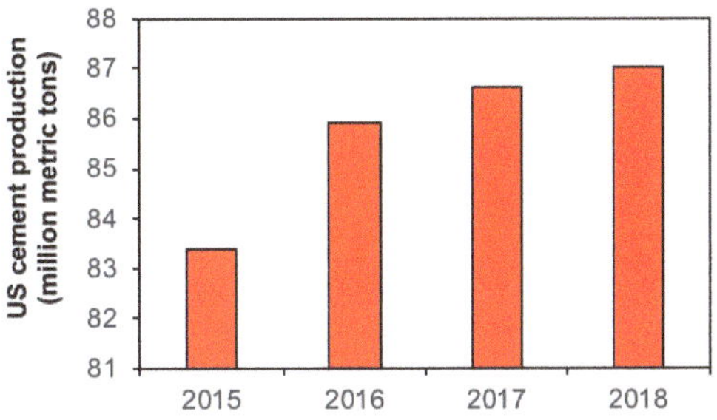

Figure 1.1: Recent US cement production.

The origins of cement date back to well over 5000 years ago when the Egyptians developed mortars composed of lime (CaO) and gypsum ($CaSO_4.2H_2O$) to hold together the enormous stone blocks of the pyramids. Three thousand years later, between 300 BC and 476 AD, the Romans developed the first durable concrete, with a cementitious matrix of lime and volcanic ash (chiefly SiO_2) from Mount Vesuvius, and used it to build the Coliseum and the huge Basilica of Constantine. The Romans also employed chemical admixtures in their cements, such as animal fat, milk, and blood, perhaps to improve the workability of their pastes. Chemicals found in these fluids are still used today to modify the setting of cement.

The use of natural cement, consisting of mixtures of lime and clay (aluminum silicates), emerged in England in the late 18th century. Joseph Aspdin (Figure 1.2) obtained the first patent on cement manufacture in 1824 (Figure 1.3). Aspdin carefully proportioned amounts of lime and clay, then pulverized the mixture and burned it in a furnace. He named his mixture 'Portland cement', because the color of the powder resembled the color of the rock quarries on the Isle of Portland, and Portland stone was the most prestigious building stone in use in England at the time. The Unites States began producing its own Portland cement in the 1870's. Technological developments such as the rotary kiln enhanced production capabilities and allowed cement to become one of the most widely used construction materials in the world.

Figure 1.2: English cement manufacturer Joseph Aspdin (1778 - 1855) who obtained the patent for Portland cement on 21 October 1824.

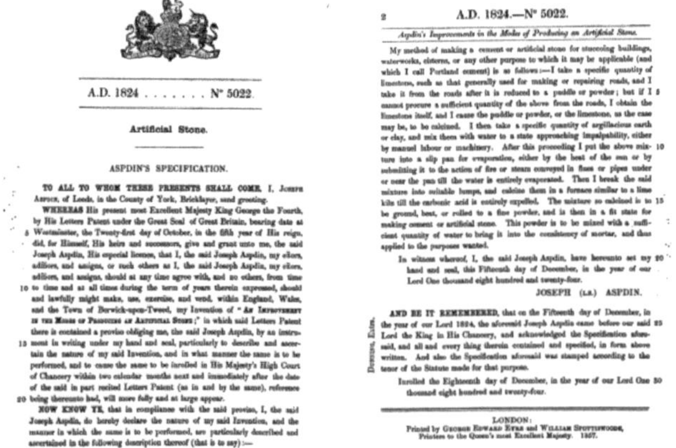

A.D. 1824 . . . . N° 5022.

Artificial Stone.

ASPDIN'S SPECIFICATION.

TO ALL TO WHOM THESE PRESENTS SHALL COME, I, Joseph Aspdin, of Leeds, in the County of York, Bricklayer, send greeting. WHEREAS His present most Excellent Majesty King George the Fourth, by His Letters Patent under the Great Seal of Great Britain, bearing date at Westminster, the Twenty-first day of October, in the fifth year of His reign, did, for Himself, His heirs and successors, give and grant unto me, the said Joseph Aspdin, His especial licence, that I, the said Joseph Aspdin, my exers, adfixors, and assigns, or such others as I, the said Joseph Aspdin, my exers, adfixors, and assigns, should at any time agree with, and no others, from time to time and at all times during the term of years therein expressed, should and lawfully might make, use, exercise, and vend, within England, Wales, and the Town of Berwick-upon-Tweed, my Invention of " AN IMPROVEMENT IN THE MODES OF PRODUCING AN ARTIFICIAL STONE;" in which said Letters Patent there is contained a proviso obliging me, the said Joseph Aspdin, by an instrument in writing under my hand and seal, particularly to describe and ascertain the nature of my said Invention, and in what manner the same is to be performed, and to cause the same to be inrolled in His Majesty's High Court of Chancery within two calendar months next and immediately after the date of the said in part recited Letters Patent (as in and by the same), reference being thereunto had, will more fully and at large appear. NOW KNOW YE, that in compliance with the said proviso, I, the said Joseph Aspdin, do hereby declare the nature of my said Invention, and the manner in which the same is to be performed, are particularly described and ascertained in the following description thereof (that is to say):—

2   A.D. 1824.—N° 5022.

Aspdin's Improvements in the Modes of Producing an Artificial Stone.

My method of making a cement or artificial stone for stuccoing buildings, waterworks, cisterns, or any other purpose to which it may be applicable (and which I call Portland cement) is as follows:—I take a specific quantity of limestone, such as that generally used for making or repairing roads, and I take it from the roads after it is reduced to a puddle or powder; but if I cannot procure a sufficient quantity of the above from the roads, I obtain the limestone itself, and I cause the puddle or powder, or the limestone, as the case may be, to be calcined. I then take a specific quantity of argillaceous earth or clay, and mix them with water to a state approaching impalpability, either by manual labour or machinery. After this proceeding I put the above mixture into a slip pan for evaporation, either by the heat of the sun or by submitting it to the action of fire or steam conveyed in flues or pipes under or near the pan till the water is entirely evaporated. Then I break the said mixture into suitable lumps, and calcine them in a furnace similar to a lime kiln till the carbonic acid is entirely expelled. The mixture so calcined is to be ground, beat, or rolled to a fine powder, and is then in a fit state for making cement or artificial stone. This powder is to be mixed with a sufficient quantity of water to bring it into the consistency of mortar, and thus applied to the purposes wanted.

In witness whereof, I, the said Joseph Aspdin, have hereunto set my hand and seal, this Fifteenth day of December, in the year of our Lord One thousand eight hundred and twenty-four.

JOSEPH (L.S.) ASPDIN.

AND BE IT REMEMBERED, that on the Fifteenth day of December, in the year of our Lord 1824, the aforesaid Joseph Aspdin came before our said Lord the King in His Chancery, and acknowledged the Specification aforesaid, and all and every thing therein contained and specified, in form above written. And also the Specification aforesaid was stamped according to the tenor of the Statute made for that purpose.

Inrolled the Eighteenth day of December, in the year of our Lord One thousand eight hundred and twenty-four.

LONDON:
Printed by GEORGE EDWARD EYRE and WILLIAM SPOTTISWOODE,
Printers to the Queen's most Excellent Majesty. 1857.

Figure 1.3: A copy of Joseph Aspdin's British Patent (BP 5022) entitled "An Improvement in the Modes of Producing an Artificial Stone" issued 21 October 1824.

Cement production may be classified by application into two primary groups: construction and energy services. The construction applications for cementing consume the lion's share of cement manufactured world-wide, but the cement produced for energy services applications is an integral part of meeting the world's energy needs and requires tighter quality control standards to meet that industry's higher demands on control of the rheological properties of the fluid slurry state, the solid state, and especially the transition from the former state to the latter, or the setting process. Applications relating to the energy services industry are the primary focus of this work. Additionally, cement may also become central to efforts in nuclear waste management by locking radioactive material within the cementitious matrix, where rates of diffusion of waste out of the cement serve as the dominant concern.

Bibliography

J. Adam, *Roman Building: Materials and Techniques*, Bloomington, Indiana, University Press (1994).

M. Gadalla, *Pyramid Illusions*, Bastet Publishing, Erie, Pennsylvania (1997).

14

Chapter 2: Application of Portland Cement in the Energy Services Industry

Portland Cement was first used in the energy services industry in 1903 to isolate the oil-containing region of the earth from down-hole water, a process modernly referred to as zonal isolation. The technique of oil well cementing was soon developed (Figure 2.1). After the primary hole is drilled, a steel casing, through which the oil will later ow, is placed inside. Drilling mud is used to assist in the actual drilling. The cement is pumped down the steel casing to the bottom of the well and then back up through the free annular space between the casing and the well, where it serves to bond the casing to the rock formation and to prevent fluids from moving from one formation to another (hence the term zonal isolation). Displacement fluids, such as fresh water, sea water, and weak acid solutions, are used to push the cement out of the casing. To avoid damage to the pumping equipment used to place the cement slurry, the cement must remain a fluid state for several hours while it is pumped into place; to avoid wasting valuable rig time, the cement should set shortly after being placed.

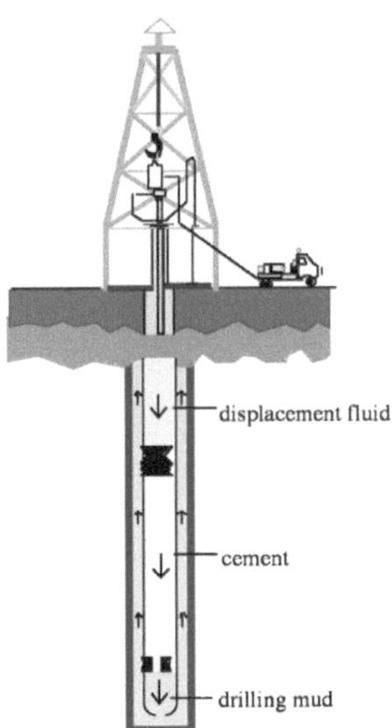

Figure 2.1: A schematic of the oil well cementing process.

In 1929, Pacific Portland Cement Company developed the first retarded cement, which allowed cement to be used in deeper oil wells than previously possible. The construction of the first bulk cement production stations in the early 1940's made the use of cement additives more practical. The desire for deeper wells, drove the development of new retarders that could predictably delay cement setting for longer periods of time than traditional retarders.

Today, oil wells are commonly 2 - 3 miles (3.2 – 4.8 km) deep, at which depth temperatures of 200 - 300 °F (93 - 148 °C) is not uncommon. Cement pumped to the bottom of these wells is subject to pressures approaching 150 MPa (21×10^3 psi). Cement must remain as thin as possible under these conditions yet begin to set and develop strength soon after it is in place so that oil drilling can continue. The demanding needs of the industry have sparked a plethora of empirical investigations into the effect of different additives on cement setting. More recently, investigations into the mechanisms of cement hydration and hydration inhibition have begun in an attempt to develop a more rational methodology for the design of cement additives.

As mentioned above, in oil well cementing it is desirable to pump a watery cement slurry down the oil well and then have this thin mixture set as quickly as possible once it is in place. Figure 2.2 shows hydration curves for regular cement hydration, and hydration with different types of additives. Several additives have been developed to control the hydration and setting of cement with varying success. In this study, reactions of four different hydration inhibitors with cement and the main mineral phases in cement have been characterized to gain insight into the different mechanisms of cement hydration inhibition and to determine which mechanisms produce more favorable cement setting behavior.

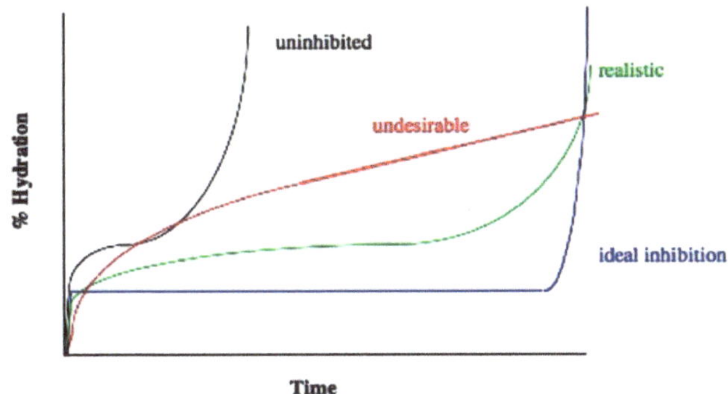

Figure 2.2: Cement hydration behavior without additives (black), with sucrose (red), with near ideal retarders such as the phosphonates (green) and the ideal behavior that is desired by the oil industry (blue).

API cement classifications

API cement is manufactured specifically to meet the needs of the oil industry. The American Petroleum Institute (API) established a set of standards that a Portland cement must meet to be considered an API cement. These standards were set so that the oil industry may obtain a product that will perform with some degree of consistency. A summary of the normal density is provided in Table 2.1.

Class	Normal density (lb/gal)
A	15.6
B	15.6
C	14.8
G	15.8
H	16.5

Table 2.1. Summary of normal density for different API classes of cement.

Class A

Intended for use from surface to 6,000 feet when special properties are not required. The properties and performance of Class A cement may be tailored with additives to meet special requirements beyond basic performance. It is similar to ASTM (American Society for Testing and Materials) Type I construction cement.

Class B

Intended for use from surface to 6,000 feet when conditions require moderate to high sulfate-resistance. Class B is similar to ASTM Type II construction cement.

Class C

Intended for use from surface to 6,000 feet when conditions require high early strength. Class C is similar to ASTM Type III cement and available in ordinary, moderate and high sulfate resistance types.

Class G

Intended for use from surface to 8,000 feet as basic cement, as manufactured, or it can be modified with additives to cover a full range of well depths and temperatures. No additions other than calcium sulfate or water, or both, shall be inter-ground and blended with the clinker during manufacture of Class G Cement. Class G cement is available in moderate and high sulfate-resistance types. Class G is similar to ASTM Type IV cements.

Class H

Intended for use from surface to 8,000 feet as basic cement, as manufactured, or it can be modified with additives to cover a full range of well depths and temperatures. No additions other than calcium sulfate or water, or both, shall be inter-ground and blended with the clinker during manufacture of Class H Cement. Available in moderate and high sulfate-resistance types. Class H is similar to ASTM Type IV cements.

Bibliography

D. K. Smith, *Cementing*, Society of Petroleum Engineers, Richardson, Texas (1990).

H. F. W. Taylor, *Cement Chemistry*, 2nd Ed., Academic Press, London (1997).

Chapter 3: The Chemical Composition of Portland Cement

There are four chief minerals present in a Portland cement grain: tricalcium silicate (Ca_3SiO_5), dicalcium silicate (Ca_2SiO_4), tricalcium aluminate ($Ca_3Al_2O_5$) and calcium aluminoferrite ($Ca_4Al_nFe_{2-n}O_7$). The formula of each of these minerals can be broken down into the basic calcium, silicon, aluminum and iron oxides (Table 3.1). Cement chemists use abbreviated nomenclature based on oxides of various elements to indicate chemical formulae of relevant species, i.e., C = CaO, S = SiO_2, A = Al_2O_3, F = Fe_2O_3. Hence, traditional cement nomenclature abbreviates each oxide as shown in Table 3.1.

Mineral	Chemical formula	Oxide composition	Abbreviation
Tricalcium silicate (alite)	Ca_3SiO_5	$3CaO.SiO_2$	C3S
Dicalcium silicate (belite)	Ca_2SiO_4	$2CaO.SiO_2$	C2S
Tricalcium aluminate	$Ca_3Al_2O_4$	$3CaO.Al_2O_3$	C3A
Tetracalcium aluminoferrite	$Ca_4Al_nFe_{2-n}O_7$	$4CaO.Al_nFe_{2-n}O_3$	C4AF

Table 3.1: Chemical formulae and cement nomenclature for major constituents of Portland cement. Abbreviation notation: C = CaO, S = SiO_2, A = Al_2O_3, F = Fe_2O_3.

The composition of cement is varied depending on the application. A typical example of cement contains 50 - 70% C3S, 15 - 30% C2S, 5 - 10% C3A, 5 - 15% C4AF, and 3 - 8% other additives or minerals (such as oxides of calcium and magnesium). It is the hydration of the calcium silicate, aluminate, and aluminoferrite minerals that causes the hardening, or setting, of cement. The ratio of C3S to C2S helps to determine how fast the cement will set, with faster setting occurring with higher C3S contents. Lower C3A content promotes resistance to sulfates. Higher amounts of ferrite lead to slower hydration. The ferrite phase causes the brownish gray color in cements, so that white cements (i.e., those that are low in C4AF) are often used for aesthetic purposes.

The calcium aluminoferrite (C4AF) forms a continuous phase around the other mineral crystallites, as the iron containing species act as a fluxing agent in the rotary kiln during cement production and are the last to solidify around the others. Figure 3.1 shows a typical cement grain.

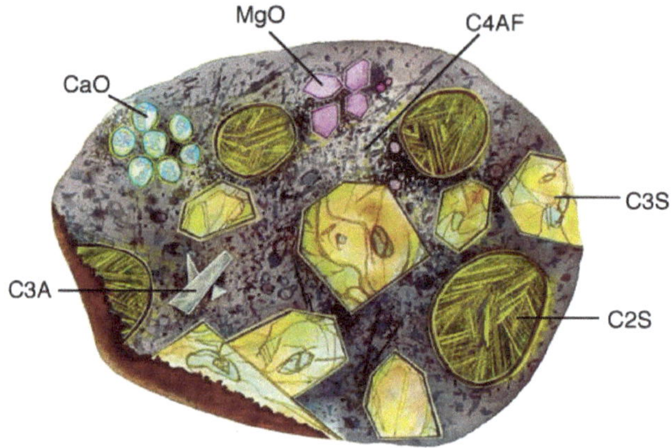

Figure 3.1: A pictorial representation of a cross-section of a cement grain. Adapted from Cement Microscopy, Halliburton Services, Duncan, OK.

It is worth noting that a given cement grain will not have the same size or even necessarily contain all the same minerals as the next grain. The heterogeneity exists not only within a given particle, but extends from grain to grain, batch-to-batch, plant to plant.

Since the early formation of C-S-H, the primary dictator of solidification and strength development, for cement depends on the quantity of C_3S available for reaction, gaining useful information about the amount of C_3S able to hydrate is of value. Since cement manufacturing is controlled primarily by regulation of silicon content and the ultimate fate of silicon is in either C_3S or C_2S, knowledge of the C_3S/C_2S ratio puts the amount of C_3S alone in the context of the whole cement. Determination of C_3S/C_2S ratios by ^{29}Si MAS NMR spectroscopy provides an effective method of analysis for cements, owing to NMR spectroscopy's direct measurement of the minerals in question without the assumptions of composition that are required for oxide analysis from X-ray fluorescence and application of the Taylor-modified Bogue equations.

Bibliography

C. L. Edwards, L. B. Alemany, and A. R. Barron, Solid state ^{29}Si NMR analysis of cements: comparing different methods of relaxation analysis for determining spin-lattice relaxation times to enable determination of the C_3S/C_2S ratio. *Ind. Eng. Chem. Res.*, 2007, **46**, 5122.

H. F. W. Taylor, Cement Chemistry, 2nd Ed., Academic Press, London (1997).

Chapter 4: Manufacture of Portland Cement

Portland Cement is manufactured by heating calcium carbonate and clay or shale in a kiln. During this process the calcium carbonate is converted to calcium oxide (also known as lime) and the clay minerals decompose to yield dicalcium silicate (Ca_2SiO_4, C2S) and other inorganic oxides such as aluminate and ferrite. Further heating melts the aluminate and ferrite phases. The lime reacts with dicalcium silicate to from tricalcium silicate (Ca_3SiO_5, C3S). As the mixture is cooled, tricalcium aluminate ($Ca_3Al_2O_6$, C3A) and tetracalcium aluminoferrite ($Ca_4Al_nFe_{2-n}O_7$, C4AF) crystallize from the melt and tricalcium silicate and the remaining dicalcium silicate undergo phase transitions. These four minerals (C3S, C2S, C3A, and C4AF) comprise the bulk of most cement mixtures. Initially Portland cement production was carried out in a furnace; however, technological developments such as the rotary kiln have enhanced production capabilities and allowed cement to become one of the most widely used construction materials.

Cement plants generally produce various grades of cement by two processes, referred to as either the wet or dry process. The dry process uses a pneumatic kiln system which uses superheated air to convert raw materials to cement, whereas the wet process slurries the raw materials in water in preparation for conversion to cement. Cement manufacturers due to its higher energy efficiency generally favor the dry process, but the wet process tends to produce cement with properties more palatable to the energy services industry. The American Petroleum Institute (API) Class H cement used in energy service applications is produced by the wet process, and thus will be the focus of the following discussion.

The cement manufacturing process begins at the quarry (Figure 4.1), where limestone formations are ripped and crushed in two crushers to a mean particle size of 4". The quarry formation is not entirely limestone, and no attempt is made to isolate the limestone from the other minerals. On the contrary, the rippers act to blend in the impurity minerals as evenly as is feasible while still maintaining an acceptable limestone content so as not to waste the formation. This is accomplished by ripping the formation face at a 45° (Figure 4.2).

Figure 4.1: Limestone quarry face in Midlothian, TX (Copyright Halliburton Energy Services).

Figure 4.2: Limestone quarry formation, showing the 45° ripping technique (Copyright Halliburton Energy Services).

The rock is quality controlled via mobile X-ray fluorescence (XRF) spectroscopy (Figure 4.3) at the starting point of a mobile covered conveyor belt system (Figure 4.4), which transports the material to a dome storage unit.

Figure 4.3: Mobile crusher and XRF unit (Copyright Halliburton Energy Services).

Figure 4.4: Covered conveyor belt for limestone transport (Copyright Halliburton Energy Services).

The dome storage unit (Figure 4.5) has a capacity of 60 kilotons and is filled by dispensing the rock from the conveyor at the top of the dome into a pile built in a circular pattern (Figure 4.6). The rock is reclaimed from storage via a raking device (Figure 4.7) that grates over the pile at the natural angle of material slide. The raked material slides to the base of the raking unit, where a second conveyor system transfers material to either of two limestone buffer bins (Figure 4.8), each of which is dedicated to a particular kiln process. There is an additional buffer bin for mill scale from a nearby steel plant, as well as a buffer bin for sand. It is worth noting at this point that the mill scale from the steel plant contains significant levels of boron, which acts as an innate retarder and seems to affect adversely, though not overly severely, the early compressive strength development when compared to cement from other plants.

Figure 4.5: Exterior view of dome storage unit with conveyor loading port (Copyright Halliburton Energy Services).

Figure 4.6: Interior view of dome storage unit, illustrating its radial piling (Copyright Halliburton Energy Services).

Figure 4.7: Raking device for reclaiming stored limestone (Copyright Halliburton Energy Services).

Figure 4.8: Buffer bin (Copyright Halliburton Energy Services).

Material leaving the buffer bins is monitored for elemental composition via XRF and feed rates are adjusted for maintaining proper ow of calcium, silicon, aluminum, and iron. The raw materials are carried to ball mills (Figure 4.9) for grinding to fine powder, which is then mixed with water. The resulting slurry is then sent to the rotary kiln for burning (Figure 4.10), or transformation into cement clinker.

Figure 4.9: Coarse grinder (Copyright Halliburton Energy Services).

Figure 4.10: A wet rotary kiln (Copyright Halliburton Energy Services).

The kilns are red to an internal material temperature of 2700 °F (1480 °C) with a fuel of finely ground coal, natural gas, and/or various waste materials (Figure 4.11). Around fifty percent of the energy expenditure in the wet kiln process is dedicated to evaporating the water from the slurry, in contrast to the dry process, which spends most of its energy on the calcining process. Since the dry process only requires approximately half the energy of the wet process, it is generally more attractive to cement manufacturers. Unfortunately, the dry process in current use produces poor API Class H cement. A fuller understanding of the differences in cement synthesis via the two processes could lead to the development of a more effective dry synthesis of Class H cements, but that is beyond the scope of this work.

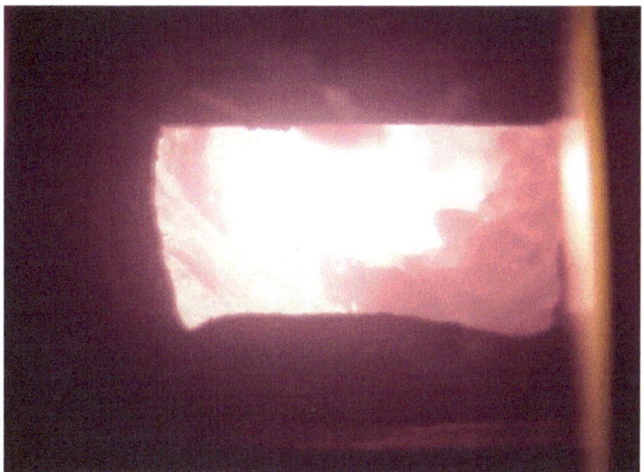

Figure 4.11: Interior view of an operational wet kiln (Copyright Halliburton Energy Services).

After the clinker leaves the kiln, it enters a cooler that uses pressurized air to cool the clinker. The energy absorbed by the air in the cooler serves to pre-heat the air for feed into the kiln. The cooled clinker is then taken to storage to await final grinding with approximately five percent gypsum by weight. After grinding to the specified fineness, the final cement powder is pneumatically transferred to storage silos until it is shipped to the customer.

Quality control of the clinker and final powder is handled via an automated X-ray diffraction/X-ray fluorescence (XRD/XRF) system, simple wet chemical analyses, simple optical microscopy, and periodic performance tests, including compressive strength and thickening time. This entire process results in the heterogeneous nanocomposite of calcium silicate and aluminate particles, among other materials, which make up a typical cement grain.

Bibliography

H. G. van Oss and A. C. Padovani, Cement manufacture and the environment: part I: chemistry and technology. *J. Ind. Ecol.*, 2002, **6**, 89.

H. G. van Oss and A. C. Padovani, Cement manufacture and the environment part II: environmental challenges and opportunities. *J. Ind. Ecol.*, 2003, 7, 93.

J. Majling and D. M. Roy, The potential of fly ash for cement manufacture. *Am. Ceram. Soc. Bull.*, 1993, **72**, 79.

P. J. Jackson in *Lea's Chemistry of Cement and Concrete*. Ed. P. Hewlett, Elsevier, New York (2003).

Chapter 5: Hydration of Portland Cement

The addition of water to dry cement powder results in a thin cement slurry that can be easily manipulated and cast into different shapes. In time, the slurry sets and develops strength through a series of hydration reactions. Hydration of cement is not linear through time, it proceeds very slowly at first, allowing the thin mixture to be properly placed before hardening. The chemical reactions that cause the delay in hardening are not completely understood; however, they are critical to developing a rational methodology for the control of cement setting.

Tri- and di-calcium silicates

The tri- and di-calcium silicates (C3S and C2S, respectively) comprise over 80% by weight of most cement. It is known that C3S is the most important phase in cement for strength development during the first month, while C2S reacts much more slowly, and contributes to the long-term strength of the cement. Both the silicate phases react with water as shown below to form calcium hydroxide and a rigid calcium-silicate hydrate gel, C-S-H,

$$2\ (CaO)_3(SiO_2)\ +\ 7\ H_2O\ \rightarrow\ (CaO)_3(SiO_2)_2 \cdot 4(H_2O)\ +\ 3\ Ca(OH)_2$$

$$2\ (CaO)_2(SiO_2)\ +\ 5\ H_2O\ \rightarrow\ (CaO)_3(SiO_2)_2 \cdot 4(H_2O)\ +\ Ca(OH)_2$$

The detailed structure of C S H is not completely known; however, it is generally agreed upon that it consists of condensed silicate tetrahedra sharing oxygen atoms with a central, calcium hydroxide-like CaO_2 layer. Calcium hydroxide consists of hexagonal layers of octahedrally coordinated calcium atoms and tetrahedrally coordinated oxygen atoms. Taylor has proposed that the structure is most similar to either tobermorite $[Ca_5Si_6O_{16}(OH)_2 \cdot 4H_2O]$ or jennite $[Ca_9Si_6O_{18}(OH)_6 \cdot 8H_2O]$, both of which share a skeletal silicate chain Figure 5.1.

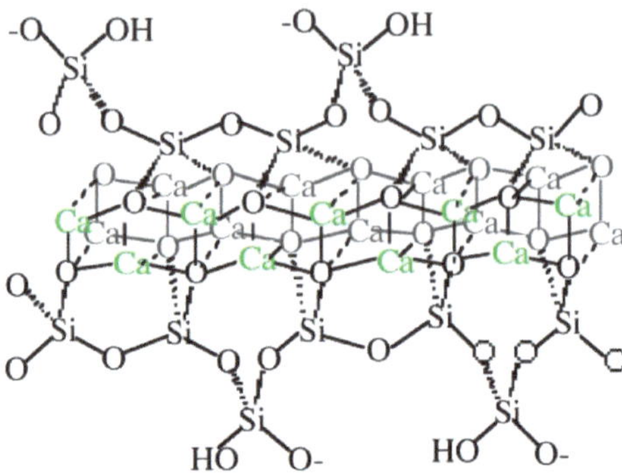

Figure 5.1: Schematic representation of tobermorite, viewed perpendicular to a polysilicate chain. Silicate ions either share oxygen atoms with a central CaO_2 core or bridge silicate tetrahedra. Interlayer calcium ions and water molecules are omitted for clarity.

Although the precise mechanism of C3S hydration is unclear, the kinetics of hydration are well known. The hydration of the calcium silicates proceeds via four distinct phases as shown in Figure 5.2. The first 15-20 minutes, termed the pre-induction period (Figure 5.2a), is marked by rapid heat evolution. During this period calcium and hydroxyl ions are released into the solution. The next, and perhaps most important, phase is the induction period (Figure 5.2b), which is characterized by very slow reactivity. During this phase, calcium oxide continues to dissolve producing a pH near 12.5. The chemical reactions that cause the induction period are not precisely known; however, it is clear that some form of an activation barrier must be overcome before hydration can continue. It has been suggested that in pure C3S, the induction period may be the length of time it takes for C-S-H to begin nucleation, which may be linked to the amount of time required for calcium ions to become supersaturated in solution. Alternatively, the induction period may be caused by the development of a small amount of an impermeable calcium-silicon-hydrate (C-S-H) gel at the surface of the particles, which slows down the migration of water to the inorganic oxides. The initial Ca/Si ratio at the surface of the particles is near 3. As calcium ions dissolve out of this C-S-H gel, the Ca/Si ratio in the gel becomes 0.8 - 1.5. This change in Ca/Si ratio corresponds to a change in gel permeability and may indicate an entirely new mechanism for C-S-H formation. As the initial C-S-H gel is transformed into the more permeable layer, hydration continues and the induction period gives way to the third phase of hydration, the acceleratory period (Figure 5.2c).

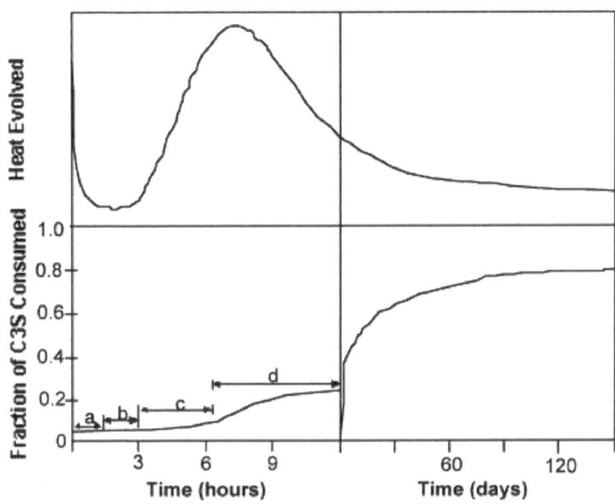

Figure 5.2: Hydration of C3S over time: (a) the pre-induction period, (b) the induction, (c) period the acceleratory period, and (d) the deceleratory period.

After ca. 3 hours of hydration, the rate of C-S-H formation increases with the amount of C-S-H formed. Solidi cation of the paste, called setting, occurs near the end of the third period. The fourth stage (Figure 5.2d) is the deceleratory period in which hydration slowly continues hardening the solid cement until the reaction is complete. The rate of hydration in this phase is determined either by the slow migration of water through C-S-H to the inner, unhydrated regions of the particles, or by the migration of H^+ through the C-S-H to the anhydrous CaO and SiO_2, and the migration of Ca^{2+} and Si^{4+} to the OH^- ions left in solution.

Calcium aluminate and ferrite

In spite of the fact that the aluminate and ferrite phases comprise less than 20% of the bulk of cement, their reactions are very important in cement and dramatically affect the hydration of the calcium silicate phase, see below. Relative to C3S, the hydration of C3A is very fast. In the absence of any additives, C3A reacts with water to form two intermediate hexagonal phases, C2AH8 and C4AH13,

$$2\ (CaO)_3(Al_2O_3) + 21\ H_2O \rightarrow (CaO)_4(Al_2O_3)\cdot13(H_2O) + (CaO)_2(Al_2O_3)\cdot8(H_2O)$$

The structure of C2AH8 is not precisely known, but C4AH13 has a layered structure based on the calcium hydroxide structure, in which one out of every three Ca^{2+} is replaced by either an Al^{3+} or Fe^{3+} with an OH^- anion in the interlayer space to balance the charge. All of the aluminum in C4AH13 is octahedral. C2AH8 and C4AH13 are meta-stable phases that spontaneously transform into the fully hydrated, thermodynamically stable cubic phase, C3AH6,

$$(CaO)_4(Al_2O_3)\cdot13(H_2O) + (CaO)_2(Al_2O_3)\cdot8(H_2O) \rightarrow 2\ (CaO)_3(Al_2O_3)\cdot6(H_2O) + 9\ H_2O$$

In C3A, aluminum coordination is tetrahedral. The structure consists of rings of aluminum tetrahedra linked through bridging oxygen atoms, which slightly distorts the aluminum environment. In C3AH6, aluminum exists as highly symmetrical, octahedral $Al(OH)_6$ units.

If the very rapid and exothermic hydration of C3A is allowed to proceed unhindered in cement, then the setting occurs too quickly, and the cement does not develop strength. Therefore, gypsum [calcium sulfate dihydrate, $CaSO_4.2(H_2O)$] is added to slow down the C3A hydration. In the presence of gypsum, tricalcium aluminate forms ettringite, $[Ca_3Al(OH)_6.12(H_2O)]_2.(SO_4)_3.2(H_2O)$,

$$(CaO)_3(Al_2O_3) + 3\ CaSO_4\cdot2(H_2O) + 26\ H_2O \rightarrow (CaO)_3(Al_2O_3)(CaSO_4)_3\cdot32(H_2O)$$

which can also be written as $C3A.3(CaSO_4).32(H_2O)$. Ettringite grows as columns of calcium, aluminum and oxygen surrounded by water and sulfate ions, as shown in Figure 5.3.

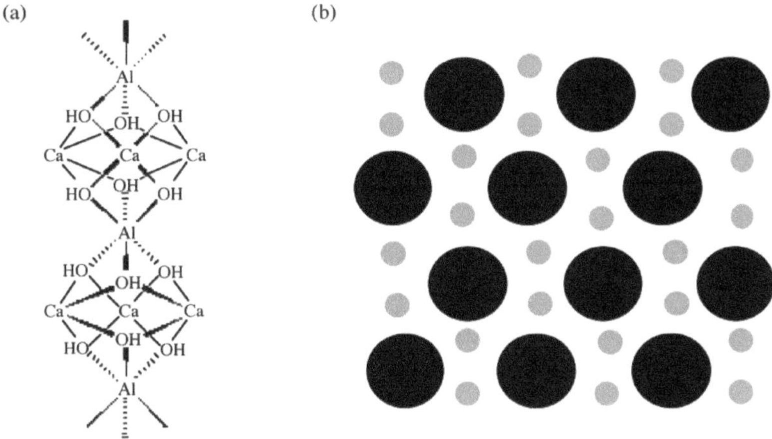

Figure 5.3: Ettringite columns (a) consisting of octahedral aluminum, tetrahedral oxygen, and 8-coordinate calcium. The coordination sphere of each calcium is filled by water and sulfate ions. The packing of the columns (b) represented by large circles, the smaller circles represent channels containing with water and sulfate ions.

Tetracalcium aluminoferrite (C4AF) reacts much like C3A, i.e., forming ettringite in the presence of gypsum. However, hydration the ferrite phase is much slower than hydration of C3A, and water is observed to bead-up on the surface of C4AF particles. This may be due to the fact that iron is not as free to migrate in the pastes as aluminum, which may cause the formation of a less permeable iron rich layer at the surface of the C4AF particles and isolated regions of iron hydroxide. In cement, if there is insufficient gypsum to convert all of the C4AF to ettringite, then an iron-rich gel forms at the surface of the silicate particles which is proposed to slow down their hydration.

Portland cement

The hydration of cement is obviously far more complex than the sum of the hydration reactions of the individual minerals. The typical depiction of a cement grain involves larger silicate particles surrounded by the much smaller C3A and C4AF particles. The setting (hydration) of cement can be broken down into several distinct periods. The more reactive aluminate and ferrite phases react first, and these reactions dramatically affect the hydration of the silicate phase. Scrivener and Pratt used TEM to develop the widely accepted model depicted in Figure 5.4.

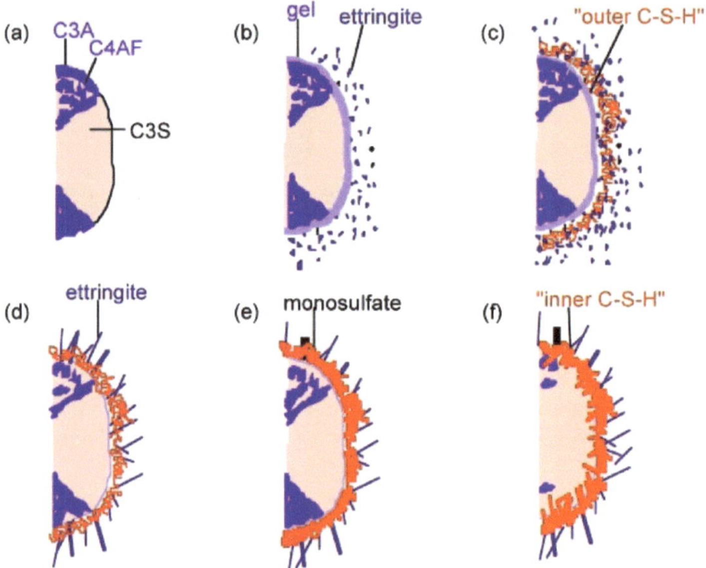

Figure 5.4: Schematic representation of anhydrous cement (a) and the effect of hydration after (b) 10 minutes, (c) 10 hours, (d) 18 hours, (e) 1 3 days, and (f) 2 weeks. Adapted from M. Bishop, PhD Thesis, Rice University (2001).

In the first few minutes of hydration (Figure 5.4b), the aluminum and iron phases react with gypsum to form an amorphous gel at the surface of the cement grains and short rods of ettringite grow. After this initial period of reactivity, cement hydration slows down and the induction period begins. After about 3 hours of hydration, the induction period ends, and the acceleratory period begins. During the period from 3 to 24 hours, about 30% of cement reacts to form calcium hydroxide and C S H. The development of C-S-H in this period occurs in 2 phases. After ca. 10 hours hydration (Figure 5.4c), C3S has produced outer C-S-H, which grows out from the ettringite rods rather than directly out from the surface of the C3S particles. Therefore, in the initial phase of the reaction, the silicate ions must migrate through the aluminum and iron rich phase to form the C-S-H. In the latter part of the acceleratory period, after 18 hours of hydration, C3A continues to react with gypsum, forming longer ettringite rods (Figure 5.4d). This network of ettringite and C-S-H appears to form a hydrating shell about 1 µm from the surface of anhydrous C3S. A small amount of inner C-S-H forms inside this shell. After 1 3 days of hydration, reactions slow down and the deceleratory period begins (Figure 5.4e). C3A reacts with ettringite to form some monosulfate. Inner C-S-H continues to grow near the C3S surface, narrowing the 1 µm gap between the hydrating shell and anhydrous C3S. The rate of hydration is likely to depend on the diffusion rate of water or ions to the anhydrous surface. After 2 weeks hydration (Figure 5.4f), the gap between the hydrating shell and the grain is completely filled with C-S-H. The original, outer C-S-H becomes more fibrous.

The chemical treatment of Class C fly ash with a 0.2 wt% of $CaCO_3$ solution results in dramatic increase in setting time and superior stability during the induction period for cement slurries at temperature conditions relevant for down-hole applications. The performance of the $CaCO_3$ treatment appears to be a consequence of each fly ash particle being provided with the necessary Ca^{2+} prior to hydration without being dependent on cement-fly ash particle interactions. Thus, the setting process is no longer dependent on the particle shape, packing and dispersion of the fly ash and cement. The observation that with the extended setting times, the induction period (Figure 5.5) follows almost ideal shape (Figure 2.2) is most important for oil well applications where it is necessary to pump a specific viscosity slurry over large distances within a specific time, without changes in viscosity that would affect pump rates.

Various attempts in the past have been made to correlate engineering performance properties with more conventional spectroscopic analyses. Barnes and co-workers were able to identify several key factors to the role of gypsum in preventing "flash set" by means of time-resolved in situ X-ray diffraction. Different tests have been applied, including, quantitative X-ray diffraction and conduction calorimetry, to attempt correlations with degree of hydration, study of additive response and its correlations with SEM, and infrared and Raman spectroscopies have been used to monitor cement hydration behavior.

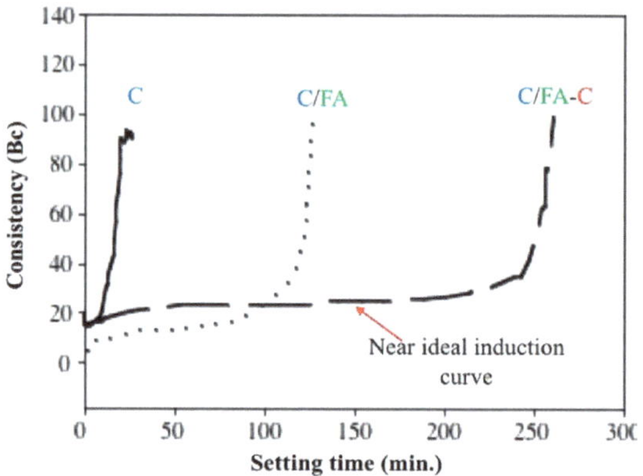

Figure 5.5: Plot of consistency (Bc) measured at 140 °F as a function of time (min.) for Class A Texas Lehigh cement (C), cement/fly ash mixture (C/FA) and cement mixed with CaCO₃ treated fly ash (C/FA-C). Adapted from C. Lupu, K. L. Jackson, S. Bard, I. Rusakova, and A. R. Barron, Control over cement setting through the use of chemically modified fly ash. *Adv. Eng. Mater.*, 2006, 8, 576. Copyright: Wiley-VCH (2006).

Determination of C_3S/C_2S ratios by ^{29}Si MAS NMR provides an effective method of analysis for cements, owing to NMR's direct measurement of the minerals in question, i.e., C_3S and C_2S. NMR ratios have demonstrated great predictive power for the determination of engineering performance properties. This is especially the case for prediction of strength development (Figure 5.6); in keeping with generally accepted understanding of cement hydration behavior, the strength development correlates with increasing C_3S content, decreasing C_2S content, and/or C_3S/C_2S ratio.

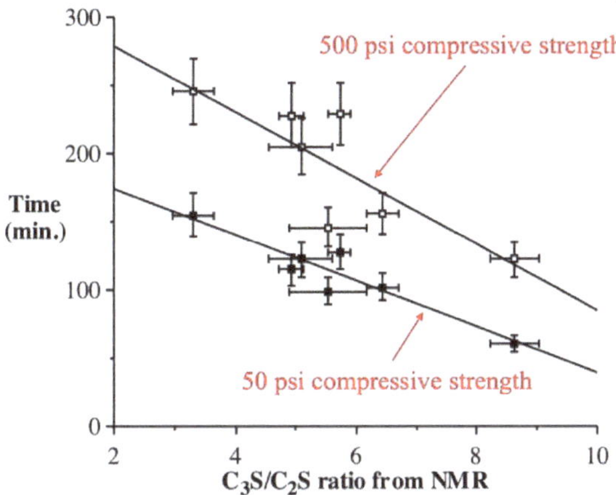

Figure 5.6: Plot of time for compressive strength development (min.) as a function of C₃S/C₂S ratio as determined from ^{29}Si MAS NMR. Adapted from C. L. Edwards, R. Morgan, L. Norman, and A. R. Barron, Correlation of cement performance property measurements with C₃S/C₂S ratio determined by solid state ^{29}Si NMR measurements. *Ind. Eng. Chem. Res.*, 2008, 47, 5456. Copyright: American Chemical Society (2008).

Bibliography

C. L. Edwards, R. Morgan, L. Norman, and A. R. Barron, Correlation of cement performance property measurements with C₃S/C₂S ratio determined by solid state ^{29}Si NMR measurements. *Ind. Eng. Chem. Res.*, 2008, **47**, 5456.

S. Gauffinet-Garrault, in *Understanding the Rheology of Concrete*, Ed. N. Roussel, Woodhead Publishing, Sawston, UK (2012).

M. Grutzeck, S. Kwan, J. Thompson, and A. Benesi, A sorosilicate model for calcium silicate hydrate (C-S-H). *J. Mater. Sci. Lett.*, 1999, **18**, 217.

C. Lupu, K. L. Jackson, S. Bard, and A. R. Barron, Water, acid and calcium carbonate pretreatment of fly ash: the effect on setting of cement-fly ash mixtures. *Ind. Eng. Chem. Res.*, 2007, **46**, 8018.

C. Lupu, K. L. Jackson, S. Bard, I. Rusakova, and A. R. Barron, Control over cement setting through the use of chemically modified fly ash. *Adv. Eng. Mater.*, 2006, **8**, 576.

V. S. Ramachandran, *Concrete Admixtures Handbook*, 2nd Edition, Noyes Publications, New Jersey (1995).

V. S. Ramanchandran, R. F. Feldman, and J. J. Beaudoin, *Concrete Science*, Heyden and Son Ltd., Philadelphia, PA (1981).

K. L. Scrivener and P. L. Pratt, Characterisation of Portland cement hydration by electron optical techniques. *Mater. Res. Soc. Symp. Proc.*, 1984, **31**, 351.

H. N. Stein and J. Stevels, Influence of silica on the hydration of 3 $CaO.SiO_2$. *J. App. Chem.*, 1964, **14**, 338.

H. F. W. Taylor, *Cement Chemistry*, 2nd Ed., Academic Press, London (1997).

H. F. W. Taylor, Proposed structure for calcium silicate hydrate gel. *J. Am. Ceram. Soc.*, 1986, **69**, 464.

Chapter 6: Hydration Inhibition of Portland Cement

In the oil industry, Portland cement supports boreholes of ever-increasing depth. This application requires a high degree of control over the setting kinetics to allow the cement to be pumped down in a liquid form. A number of chemical inhibitors are employed to delay the setting time. The ideal inhibitor for oil well cementing would predictably delay the setting of cement, and then suddenly allow hydration to continue at a rapid rate.

A wide range of compounds show set inhibition of the hydration of Portland cement. Some common examples include, sucrose, tartaric acid, gluconic acid δ-lactone, lignosulfonate, and organic phosphonic acids, in particular nitrilo-*tris*(methylene)phosphonic acid (H$_6$ntmp). The structures of these retarders are shown in Figure 6.1.

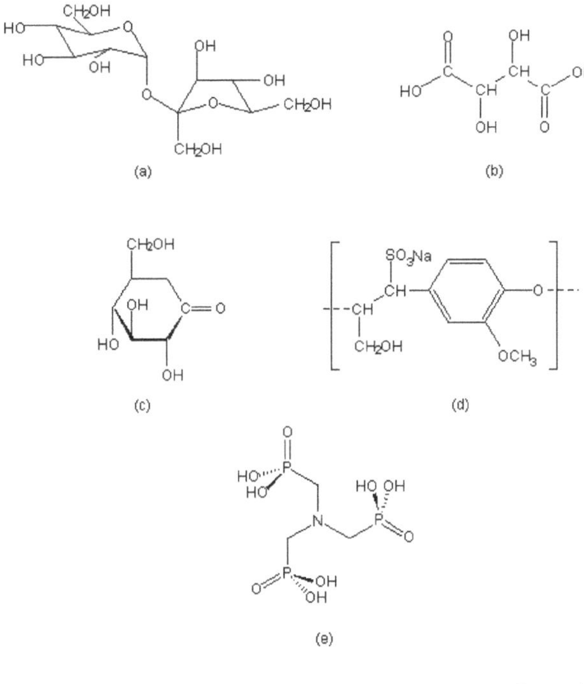

Figure 6.1: Structural formulae of common cement retarders. (a) sucrose, (b) tartaric acid, (c) gluconic acid d-lactone, (d) sodium lignosulfonate, and (e) nitrilo-tris(methylene)phosphonic acid (H$_6$ntmp).

In spite of the fact that the science of cement hydration inhibition has been investigated for over 40 years, the mechanistic details are still the subject of much speculation. There are five primary models for cement hydration inhibition: calcium complexation, nucleation poisoning, surface adsorption, protective coating/osmotic bursting, and dissolution-precipitation. A summary of the characteristic behavior of selected retarders is shown in Table 6.1.

Retarder	Characteristic behavior
Sucrose	Ca binding, acts directly on silicates, accelerates ettringite formation
Tartaric acid	Acts via calcium complexation and calcium tartrate coating, inhibits ettringite formation
Lignosulfonate	Accelerates ettringite formation, calcium becomes incorporated into the polymer matrix during hydration, forms a diffusion barrier
Nitrilo-*tris*(methylene)phosphonic acid (H_6ntmp)	Promotes Ca dissolution, forms [Ca(H_6ntmp)], heterogeneous nucleation on aluminates creates a protective coating around the grain

Table 6.1: Summary of the behavior of various hydration retarders.

Calcium complexation

Inhibition by calcium complexation relies largely on the requirement that small calcium oxide/hydroxide templates must form in the pore water of cement pastes before silicate tetrahedra can condense into dimeric and oligomeric silicates to form C S H. Calcium complexation involves either removing calcium from solution by forming insoluble salts, or chelating calcium in solution. Calcium complexation lowers the amount of calcium effectively in solution, delaying the time to $Ca(OH)_2$ super-saturation and preventing precipitation of the necessary templates. Simple calcium complexation should dramatically increase the amount of $Si(OH)_4$ tetrahedra in solution, and indeed this is observed with most retarders. However, if the retarder were acting solely by calcium complexation, then one molecule of retarder would be required per calcium ion in solution, and good inhibitors are used in much smaller quantities, on the order of 0.1-2% by weight of cement. In addition, there is no simple correlation between either calcium binding strength or calcium salt solubility and retarding ability. Yet it has been shown that in pure systems, i.e., of C3S and C2S, that the lime concentration in solutions is the most important factor in determining the precipitation of C-S-H. Therefore, although calcium complexation must play some role in inhibition, other mechanisms of inhibition must be at work as well. An example of a retarder that operates primarily through calcium complexation is tartaric acid, however, the formation of insoluble calcium tartrate on cement grains suggest that dissolution/precipitation occurs in addition (Figure 6.2).

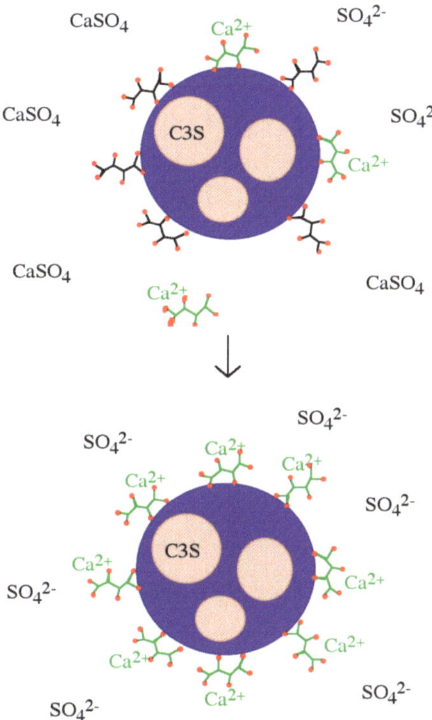

Figure 6.2: Schematic representation of cement hydration in the presence of tartaric acid. The aluminate (and aluminate ferrate) phases (shown in purple) surround the silicate phases (C3S and C2S). Tartaric acid adsorbs onto the aluminum surfaces and reacts with calcium ions from gypsum to deposit a thick calcium tartrate coating on the cement grain. Adapted from M. Bishop, PhD Thesis, Rice University, 2001.

Nucleation poisoning

As with calcium complexation, nucleation poisoning must rely on the formation of small calcium ox- ide/hydroxide templates in the pore water of cement pastes before silicate tetrahedra can condense into dimeric and oligomeric silicates to form C-S-H. Inhibition by nucleation poisoning is where the retarder blocks the growth of C-S-H or $Ca(OH)_2$ crystals through inhibiting agglomerates of calcium ions from forming the necessary hexagonal pattern. Nucleation inhibitors act on the surface of small clusters, therefore, less than one molecule of retarder per calcium ion is required to produce dramatic results. This type of inhibition also results in an increase in the amount of silicate ions in solution, as condensation of silicate chains onto calcium oxide templates to form the C-S-H is inhibited. As a retarder sucrose acts via nucleation poisoning/surface adsorption.

Surface adsorption

Surface adsorption of inhibitors directly onto the surface of either the anhydrous or (more likely) the partially hydrated mineral surfaces blocks future reactions with water. In addition, if such inhibitors are large and anionic, then they produce a negative charge at the surface of the cement grains, causing the grains to repel each other thereby reducing inter-particle interactions. Lignosulfonates are typical of retarders that act via surface adsorption.

Protective coating/osmotic bursting

The formation of a protective coating with its subsequent bursting due to the build-up of osmotic pressure was originally posited to explain the existence of the induction period in C3S and cement hydration, however it may be applied to inhibition in general. In this mechanism, a semi-permeable layer at the surface of the cement grain forms and slows down the migration of water and lengthens the induction period. Osmosis will drive water through the semi-permeable membrane towards the un-hydrated mineral, and eventually the ow of water creates higher pressure inside the protective coating and the layer bursts. Hydration is then allowed to continue at a normal rate.

Dissolution-precipitation

A detailed study of several retarders (in particular the organic phophonates) has shown that the actually accelerate certain stages of the hydration process. This is unexpected since the phosphonates have been termed super retarders, due to their increased effect on cement hydration relative to the effect of conventional retarders. So how is it that a retarder can be so efficient at hydration inhibition at the same time as accelerating the process? The ability of phosphonates to retard cement setting is due to the lengthening the induction period, without slowing down the time it takes for setting to occur (once the acceleratory period has begun).

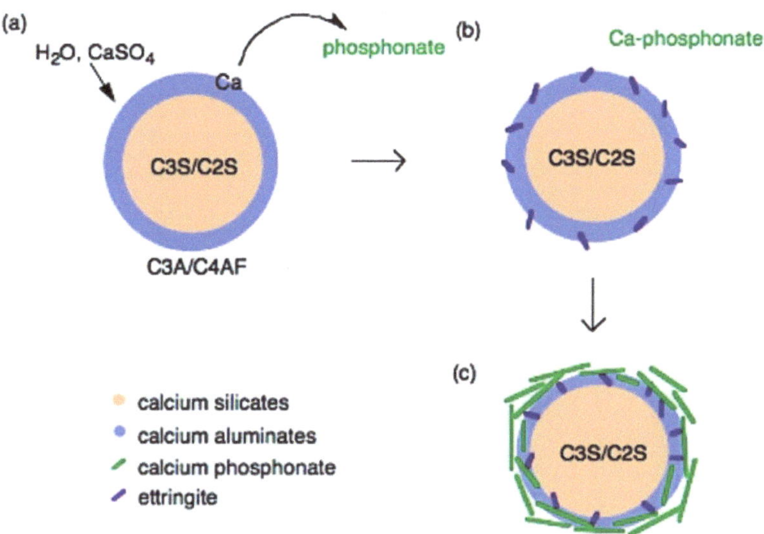

Figure 6.3: Schematic representation of the H₆ntmp inhibition of cement showing (a) the phosphonic acid promoting calcium dissolution, allowing water and gypsum to react with the aluminum phases at the surface of the cement grain, (b) the formation of a meta-stable calcium phosphonate, which precipitates onto the hydrating aluminate surfaces (c), forming a barrier to water and sulfate diffusion. Adapted from M. Bishop, PhD Thesis, Rice University, 2001.

Phosphonates are known to complex calcium and other M^{2+} cations, poison the nucleation and growth of barium sulfate crystals, and inhibit the hydration of Fe_2O_3 and Al_2O_3 surfaces via direct surface adsorption, thus it was assumed that with regard to cement hydration inhibition occurred by one of these mechanisms. However, the mechanism by which phosphonates inhibit cement hydration consists of two steps. First dissolution, whereby calcium is extracted from the surface of the cement grains (Figure 6.3a) exposing the aluminum rich surface to enhanced (catalyzes) hydration and ettringite formation (Figure 6.3b). Second precipitation, whereby the soluble calcium-phosphonate oligomerizes either in solution or on the hydrate surface to form an insoluble polymeric Ca-phosphonate (Figure 6.3c). The Ca-phosphonate material binds to the surface of the cement grains inhibiting further hydration by acting as a diffusion barrier to water as well as a nucleation inhibitor.

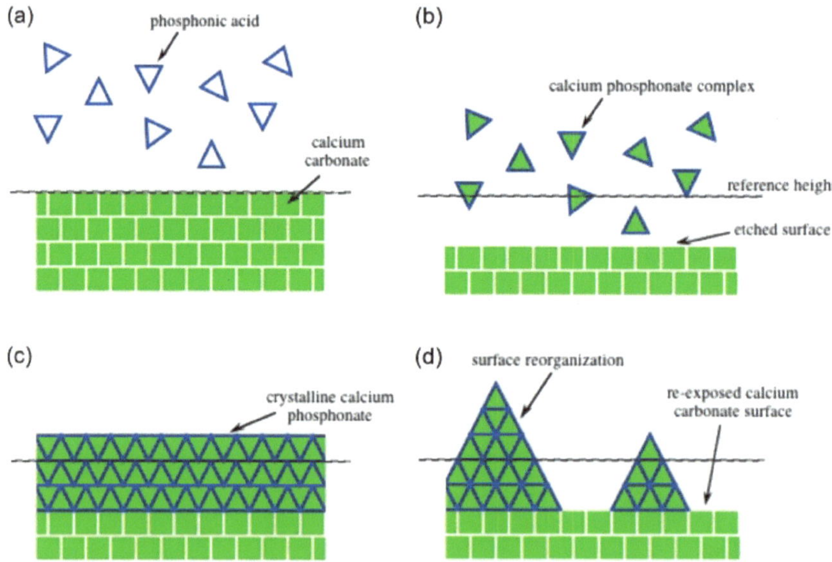

Figure 6.4. Schematic representation of the reaction between H₆ntmp with the calcite surface. Adapted from C, Lupu, R. S. Arvidson, A. Lüttge and A. R. Barron, Phosphonate mediated surface reaction and reorganization: implications for the mechanism controlling cement hydration inhibition. *Chem. Commun.***, 2005, 2354. Copyright: Royal Society of Chemistry (2005).**

Based upon vertical scanning interferometry (VLS) and X-ray photoelectron spectroscopic (XPS) data for the reaction of H_6ntmp with the surface of $CaCO_3$, it has been confirmed that a restructuring of the calcium phosphonate surface occurs exposing unreacted calcite. A schematic representation of this process is shown in Figure 6.4.

Bibliography

M. Bishop and A. R. Barron, Cement Hydration Inhibition with Sucrose, Tartaric Acid, and Lignosulfonate: Analytical and Spectroscopic Study. *Ind. Eng. Chem. Res.*, 2006, **45**, 7042.

M. Bishop, S. G. Bott, and A. R. Barron, A new mechanism for cement hydration inhibition: solid-state chemistry of calcium nitrilotris(methylene)triphosphonate. *Chem. Mater.*, 2003, **15**, 3074.

D. Double, New developments in understanding the chemistry of cement hydration. *Phil. Trans. R. Soc. London*, 1983, **A30**, 53.

C, Lupu, R. S. Arvidson, A. Lüttge and A. R. Barron, Phosphonate mediated surface reaction and reorganization: implications for the mechanism controlling cement hydration inhibition. *Chem. Commun.*, 2005, 2354.

N. Thomas and J. Birchall, The retarding action of sugars on cement hydration. *Cement Concrete Res.*, 1983, **13**, 830.